CARTHAGINIANS: SHIPPING AND TRADING LESSONS FROM HISTORY

By: Mustafa Nejem

CONTENTS

Introduction

Carthage is a name that remains relatively unknown to many. Even those familiar with the Phoenicians may not fully grasp the depth of Carthage's contributions to ancient maritime and trade history. Much like their Phoenician ancestors and other contemporary civilisations, Carthaginians were interested in exploration and navigation at sea, establishing a successful trading empire and developing impressive shipbuilding skills. Their influence in these domains is a remarkable yet often overlooked aspect of ancient history.

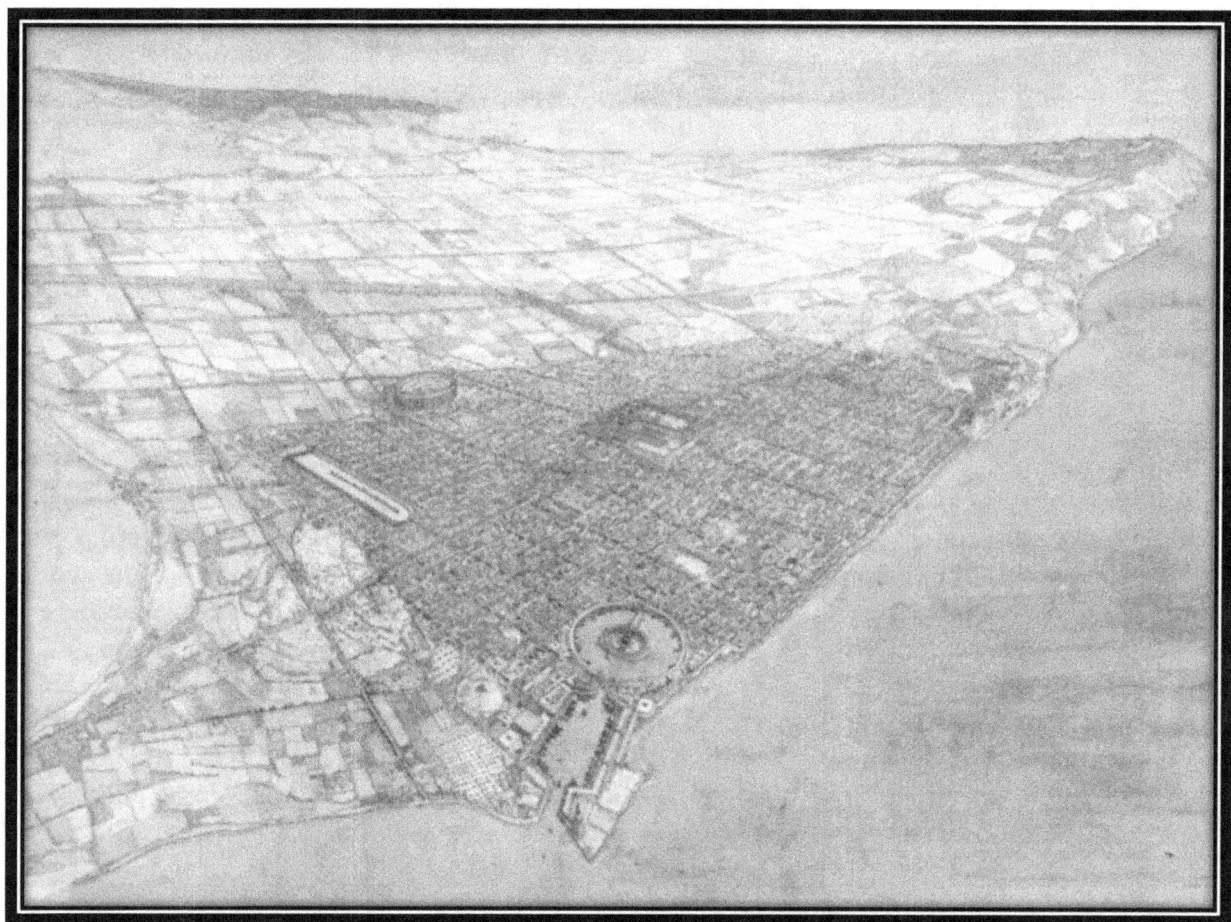

First, let's delve into Carthage's shipbuilding skills, which played a significant role in their numerous voyages and exchange of valuable goods. The Carthaginians were renowned for their remarkable shipbuilding abilities. They crafted fast and durable ships, employing innovative materials such as wood and bronze, a noteworthy feat for their era. These vessels were versatile, serving both trade and military purposes. Carthaginian shipbuilding left a profound mark on trade, enabling the efficient transport of goods and contributing to economic growth.

Certainly, with its enhanced shipbuilding skills, Carthage faced threats from neighbouring cities due to its prominent geographical advantages. In response, they bolstered their naval power. The Carthaginians possessed a formidable navy, particularly evident during the Punic Wars, and operated well-designed warships crewed by skilled sailors. Their naval prowess extended to tactics like boarding actions, coastal raids, and the employment of mercenaries.

Carthage's naval power played a pivotal role in controlling trade routes in the Mediterranean, ensuring the safety and stability of trade in the Mediterranean Sea.

When it comes to trading, the Carthaginians excelled and dominated their era through remarkable marketing and trading strategies. Their strategic location in modern-day Tunisia enabled them to establish a vast trade network at the crossroads of the Mediterranean. They diversified their goods, offering fine clothing, slaves, and exotic items. Furthermore, ethical trade practices and strong relationships with partners were pivotal for Carthaginian traders. Their culture influenced trade choices and the production of high-quality goods, ensuring long-term success and profitability.

These high-quality goods included precious metals like gold, silver, tin, and more. They also traded spices and foodstuffs from various regions, highlighting their ability to diversify their trade. The Carthaginians specialised in crafting fine textiles and the famous purple dye. Additionally, they traded ivory, animal hides, and a range of other merchandise, including slaves, weapons, tools, and exquisite art pieces. Their specialisation in specific categories, their craftsmanship, the diversity of goods they dealt with, and their unwavering commitment to high-quality products all contributed to their remarkable success.

Likewise, these high-quality goods also helped them foster strong trading partnerships. They established trade relationships across Europe, Asia, and Africa, securing vital resources and trade goods while expanding their influence. By actively nurturing these alliances, Carthage leveraged diplomatic expertise alongside its trade prowess. Notable alliances included ancient Egypt and fellow Phoenician city-states, showcasing their goodwill and diplomatic acumen.

In the same way, the Carthaginians cultivated these trading partners while establishing essential trade routes and colonies. They created an extensive network of trade routes connecting their ports and colonies, facilitating the exchange of valuable goods from around the ancient world. Carthage had major trade routes encompassing the Mediterranean, Trans-Saharan, and potential exploratory routes in the Atlantic. With its prime location, the city of Carthage played a pivotal role as a bustling port and naval power. The economic impacts of the Carthaginian trade were substantial, leading to manufacturing growth, agricultural expansion, and wealth accumulation. The chapter also highlights the remarkable voyages of Carthaginian leader Hanno and his encounter with the "gorillas," possibly early records of great apes. This demonstrates how adventurous and daring the Carthaginians were in their pursuit of trade and exploration.

The minting of coins was another integral aspect of their trade, enabling them to maintain a balanced trade system. Carthage utilised a variety of metals, including gold, silver, and bronze, to create a wide range of coins suitable for different transactions. These coins featured intricate designs and inscriptions, often reflecting Carthaginian cultural and religious traditions. Their currency system drew influence from Greek and Egyptian cultures, showcasing the rich tapestry of Carthaginian coinage. The wide acceptance of Carthaginian coins in the Mediterranean greatly facilitated trade, playing a pivotal role in Carthage's prosperity and economic influence.

Ultimately, we will explore how the Carthaginian trade left a deep mark on the ancient world, shaping economic landscapes and fostering cultural exchange and wealth accumulation. The prosperity that resulted from their extensive trading networks was not confined to their civilisation; it spread across the Mediterranean like ripples in a pond.

The economic impact of the Carthaginian trade was profound. Their ability to connect the Western and Eastern Mediterranean regions made them a critical bridge between civilisations. Goods, ideas, and cultures flowed along these trade routes, creating a rich tapestry of influences and interactions. This cultural exchange was a two-way street, with Carthaginian traders bringing their wares and absorbing other nations' knowledge and products. The result was an amalgamation of diverse elements, enriching the cultures and societies involved. Furthermore, this exchange often led to innovations and improvements in various industries, further driving economic growth.

This book will delve into the intricate web of trade routes, the treasures that travelled along them, and the profound impact that the Carthaginian trade network had on the ancient Mediterranean world. The legacy of their trading success is a testament to the transformative power of commerce and exchange on both a regional and global scale.

CHAPTER 1

CARTHAGINIAN'S SHIPBUILDING SKILLS

The Carthaginians were well known for their outstanding shipbuilding skills. Their ship designs include good speed, manoeuvrability, and striking power. Carthaginian shipbuilders have employed many innovative shipbuilding materials for corrosion resistance and durability. All these vessels were crafted from sturdy woods, bronze, and other notable materials. Their ships were known for navigating waterways from open seas to coastal regions. This versatility made the Carthaginian ships valuable for naval warfare and trade. This shipbuilding legacy of the ancient civilisation continues to inspire modern maritime practices and technological advancements.

Carthaginian Ship Design

Warships, including Quinquereme, offer very remarkable designs. All these vessels were long and narrow, which made them highly suited for speed and operation. This design of the ships allowed them to navigate through the Mediterranean region's narrow and shallow coastal waters. The most famous Carthaginian warship was Quinquereme. It was made with five rows of oars. This arrangement of oars style made it formidable in Naval battles. The front of these Carthaginian warships was made of bronze with ramming structures. This helped the ships to ram in the enemy shapes, which caused much damage to the opponents and caused them to sink.

Carthaginians have used numerous materials for the construction of their ships. The primary material that was used for the framework was wood. This wood is comprised of oak, cedar, or pine. These woods were chosen because of their strength and resistance to rotting in salt water. They also used bronze in the ship construction. Bronze was used for a range of components, including fittings, prow, and nails. Bronze corrosion resistance was also guaranteed and crucial for the ships' longevity and guarantee. The sails were made from papyrus or linen, which helped provide flexibility and durability for sailing. The ship's rigging included material which was made from flax or hemp. It allowed for the efficient control of sails.

To make the hull seem waterproof, Carthaginians used pitch or tar material. This helped in the prevention of water from moving into the ship structure. They have also used some elements for decorating the ships with intricate carvings and decorative elements. All these elements showcase their craftsmanship and aesthetic appeal.

So, the combination of all these materials used in the construction of the Carthaginian ships resulted in warships that were powerful, well-constructed, and good in design. These ships were capable of withstanding the rigours of the Mediterranean Sea. They were also very powerful for any battle because they resulted from the skills of Carthaginian shipbuilders.

Construction of Ships by Carthaginians

The process of ship construction by Carthaginians was complex and labour-inclusive. Following is a detailed description of the ships' construction process and design. The shipbuilding process begins with complete planning and details. The naval architects and other craftsmen created blueprints and specifications for the ship before construction. While creating the ship's blueprint, they consider its size, shape, purpose, and capabilities. The construction of these started by laying the sturdy kill. It served as the ship's backbone, which was positioned on a cradle that allowed the ship to be built vertically.

After that, the ribs, beams, and wooden frames were attached to the kill. These accessories formed the ship's base, providing its structure and shape. This frame was carefully constructed to ensure the durability and stability of the ship. The next step was to create the planks. The wooden planks were then fixed with the framework to balance the ship and create the ship's hull.

The planks were also secured using joints. Bronze, the key material in Carthaginian shipbuilding, was used for numerous ship components. These components include rams, nails, and structural elements. These bronze bolts and nails were attached to the wooden frame to strengthen the ship's construction. Then, rigging for the ship was done through hemp or flex, which allowed for the efficient control of sails. Next, waterproofing was done to prevent water infiltration and protect the water from sinking out of the ship.

Once the Carthaginians constructed the ship, it went through different sea trials and testing to ensure the performance and safety standards of the ship. After successful testing, the ship was also launched into the sea. So, constructing Carthaginian ships was a highly skilled task with the collaborative effort of craftsmen and shipbuilders. All these advanced techniques of shipbuilders allowed the Carthaginian vessels to dominate in the Mediterranean Sea and play a very important role in naval history.

Innovations in Shipbuilding Techniques by Carthaginian

Carthaginians have made numerous innovations and advancements in the field of shipbuilding. This made them more prominent in the ancient world. They were pioneers in utilising bronze for making different ship components. The Quinquereme design was also a significant innovation. This design provided superior speed and rowing power. These ships also had a shallow draft, which helped them navigate both open and shallow waters.

Carthaginians also introduced bronze prow for ramming into the enemies' vessels. They made numerous designs and carvings on the ships meant for warfare or trade. All these innovations in the shipbuilding techniques allowed the Carthaginian ships to be more prominent in the Mediterranean Sea. They played a vital role in the Punic Wars and other battles.

Notable Carthaginian Ships

Some notable Carthaginian ships used in the Navel history comprise advanced designs and purposes.

Marsala Ship:

Marsala ship was found in Marsala, Sicily. It was believed to date back to the first Punic War. This ship was considered a prime example of a unique warship. The ship's design included bronze-made ramming, prow which indicated the ship's role as a warship. It was used in numerous battles in the ancient history. The bronze rams were used for ramming enemy vessels. This ship was very authentic and notable in different wars. The construction was also advanced because of the Carthaginian craftsmanship.

Mazara del Vallo Wreck:

This ship was discovered near Mazara del Vallo, Sicily. It was attributed to the Punic period. This ship was designed for merchants and traders to exchange numerous commodities. The ship comprised amphoras, which shows that this ship was meant for trading. Its operation as a trading vessel involved the transport of goods by Carthaginians.

Motya Ship:

Motya Ship was another well-known ship of the Carthaginian times. It was located in Motya, Sicily. This shipwreck was discovered in the 6th century BCE. It was considered as a fishing or merchant vessel. Moreover, its design includes a simple and open hull. This ship was equipped for trading and fishing purposes. The most remarkable aspect of this ship was its cargo. The Amphoras were used, which were essential for the trading network. It also highlights the ancient maritime history.

Giglio Shipwreck:

This shipwreck was discovered in Giglio, Italy, in the third century BCE. This shipwreck was used for different activities in the sea. It included amphoras and other cargo items, indicating that the ship was used for trading purposes. It is a very important archaeological discovery by Carthaginians, offering insight into the ancient Maritime trade. It also contributes to understanding cultural exchanges and the type of goods in the ancient world. This shipwreck also represents a valuable archaeological insight and has great historical significance in the Mediterranean region.

Kerkouane Ships:

These ships were located in the underwater site of Kerkouane, Tunisia. It was discovered in the 1920s and is believed to have been used for trading purposes by Carthaginians. They are considered the remnants of Punic ships. This artefact provides a great insight into Argentinian history, trade, and daily life activities. There were numerous maritime artefacts, mainly pottery and amphorae, for exchanging goods through these ships.

Impacts of Shipbuilding on Trading:

Shipbuilding has always had a profound impact on trading throughout history. It has been influencing the trade routes, the amount of goods transported, and the reach of the goods in the market. The advancement in shipbuilding techniques has led to the construction of more efficient and larger vessels. It has enabled the traders and the merchants to transport large quantities of goods. As a result, it reduces the cost per unit and increases the profitability of different businesses. The Carthaginians have shown in ancient times that shipbuilding greatly impacted increasing efficiency while trading.

Similarly, developing different ships with long-range capability is a global trade speciality. Ships like Carrivals were allowed for the discovery of new trade routes. They connected distant parts of the world. So, ships were used for exploration and Global trade as well. Moreover, the construction of ships helped open new trade routes and markets. The ships were innovated for navigation in the challenging waters, which expanded the geographical scope of trade. The ability of Carthaginians to build much larger ships contributed to globalisation. It has enabled the rapid exchange of culture, ideas, and goods across the continents and oceans.

In addition, shipbuilding has also been a source of economic growth in different regions. It provides many employment opportunities as well and contributes to the local economy. Shipbuilding techniques and tips had a great impact on the regional economies. Also, the movement of goods through these ships has boosted the cultural exchange, engaging the people with new products and traditions from the regions.

Shipbuilding has also driven technological innovations in different areas. These areas include navigation, safety, and propulsion, which makes the trade more efficient and safer. So, modern shipbuilding is increasingly focused on environment-friendly designs. Eco-friendly ships help in mitigating the impact of trading.

As a result, the advancements in shipbuilding design have facilitated economic growth, cultural exchange, and the movement of goods on a large scale.

Lessons for Modern Shipbuilding and Technological Advancements:
There are numerous lessons for the modern shipbuilders drawn from the Carthaginian practices. Modern businesses can incorporate all the lessons into modern shipping practices, which can lead to the development of more efficient ships. It can also increase the production of environment-friendly vessels and ships well-suited to specific conditions. All these lessons provide insight into getting benefits to the global shipping industry and maritime sustainability.

Innovative Use of Materials:
Different materials were used for innovating the ships in the past. The Carthaginians used bronze for ship construction, which demonstrated the importance of innovative materials. The Carthaginians also used the alloys to enhance the vessel's performance. So modern shipbuilders should also explore advanced materials. They should understand the importance of high-strengthening alloys and bronze. Also, they should use alloys to increase the performance of the ships and their durability.

It also provides corrosion resistance, a plus point for shipbuilders in creating vessels and trading. The innovative use of the materials can easily reduce the maintenance cost. It will also increase the profitability for the Shipbuilders of the modern businesses. As a result, this will extend the life of the ships, which will meet the sustainability goals.

Efficiency and Sustainability:
The Carthaginians have always focused on sustainability and efficiency during ship construction. It helped them in the smooth flow of work and minimise risk. Modern businesses should also employ shipbuilding that should prioritise eco-friendly designs, emission-reduction technology, and energy-efficient propulsion systems. Modern shipbuilders should innovate propulsion techniques, including alternative fuels and hybrid systems. They help in minimising the environmental impact. It also complies with the environmental regulations. So, the different shipbuilding tips and techniques can be taken from the Carthaginians for a smooth and sustainable ship design by modern businesses.

Specific Design for Specific Roles:
The Carthaginian ships were designed for a clear purpose. Either they were used for war or trade. So, modern shipbuilding should continue prioritising the specific design for specific roles. Modern businesses should consider different factors like speed, fuel efficiency, cargo capacity, and adaptability to environmental conditions. This is because the specialised designs of the ships help optimise the vessel's performance. It will have a great

impact on the operation of the ships. So, modern shipbuilders can easily increase their cargo and trading by taking this lesson of designing ships for a specific role and condition.

Integration of Technology:

The Carthaginians were very famous for the integration of technology. They have used many advanced tools and techniques. So, the modern shipbuilders can also learn from Carthaginians. There are numerous technologies, including advanced software for computer-aided designs, the application of sensors, and automated manufacturing processes, which help reduce the cost and provide a streamlined construction of ships. It will also enhance the precision and accuracy of shipbuilding while maintaining all the shipbuilding measures.

Additionally, the use of data Analytics and artificial intelligence will also help in improving the decision-making throughout the ship's construction. It will help provide great insight into the ship's life cycle from design to operation. So modern businesses should not hesitate. Rather, they should take some tips and techniques from Carthaginians and apply modern technologies in their businesses.

Incorporating the lessons into modern shipbuilding ties helps the development of ships that will be environment-friendly, efficient, and smooth to work. They will ultimately benefit the modern shipping industry and global maritime sustainability.

The Carthaginians were known for their impressive shipbuilding skills. They have excelled in constructing trading vessels and warships. They utilised advanced woodworking techniques, making the ships sturdy and efficient. Their expertise in shipbuilding played a very important role in maritime endeavours and military campaigns, including the Punic Wars. Several lessons for modern shipbuilders can also be gained from the Carthaginian's innovative construction methods. These ships were known for durability and high-efficiency operation.

By understanding the choice of materials and the craftsmanship, modern shipbuilders can also design vessels that should have longevity and enhance durability. The warships of Carthaginians were formidable In the naval battles. So, by studying the designs and tactics, modern shipbuilders can get insights for developing effective warship designs. The Carthaginian shipbuilders' strategies, tips, and techniques give a better understanding of the ship's improved designs, construction techniques, and operation.

Chapter 2

CARTHAGINIAN NAVAL POWER

The Carthaginians were known for their remarkable naval power. During ancient times, they built a very powerful navy that played an important role in the Rome conflicts, notably during the Punit War. They had well-designed warships that included Quinquereme and skilled sailors. The navy of Carthaginians was a vital component of their military strength. It also allowed them to control the important trade routes in the Mediterranean region. However, the Carthage was defeated by the Roman Navy in the Punic Wars. But after that, they enhanced their naval skills and focused on shipbuilding navigation and naval tactics.

Carthaginian Naval Skills

The Carthaginian naval skills were highly adopted during ancient times. They were known for their expertise in naval strength, which made shipbuilding easier. They constructed advanced warships that were well-designed for both combat and speed. All the ships were built with wooden planks and metal, which made them very suitable for naval engagement. They also experienced and skilled sailors. Carthage comprised a very well-trained crew responsible for operating the ships efficiently. Moreover, the Carthaginian navigators were very adaptable for maritime navigation.

They understood the Mediterranean geography well, which enabled them to control the sea routes. They employed numerous effective naval techniques for capturing the enemy ships. These techniques included boarding actions and the use of grappling hooks. These different techniques were focused on getting a victory against the enemies before every battle against the Romans. Although the Romans defeated them in ancient times, their never-ending legacy is noteworthy in ancient naval history.

The Mighty Carthaginian Navy

The Carthaginian Navy was commonly called The Mighty Carthaginian Navy, which was a remarkable maritime force in ancient history. It played a very significant role in the Carthaginian battles. It has also influenced the trade and economy of the Mediterranean region. The control of key sea routes in the Mediterranean Sea was to establish a very powerful maritime empire. They also control trade and commerce in that region. The Carthaginian sailors were trained and experienced, enabling them to operate the warships in naval engagements effectively.

A famous general named Hannibal used the Carthaginian navy to transport his army. The army included war elephants across the Alps during the Second Punic War. It was surprising and challenging for the Roman Republic to see a big army from the Carthaginian side against them.

The Carthaginian navy has played a very important role in the wars against Rome. These naval battles were notable, including the battle of Aegates Islands and the Battle of Drepana. Despite their naval progress, the Romans ultimately defeated the Carthaginians, which led to the destruction of Carthage in 146 BCE. But the legacy of The Mighty Carthaginian Navy continues. It is remembered as a significant maritime force in ancient history from which the modern world can gain numerous lessons.

Notable Naval Victories and Admirals
Carthage had numerous notable Naval victories throughout its history. There are many significant naval engagements of the Carthaginian commanders, which are the following:

Battle of Mylae:
This battle was fought in 260 BCE. It was one of the earliest Battles of the First Punic War. In this battle, the Carthaginian admiral Hannibal Gisco faced the Roman admiral Gaius Duilius. Though the struggle and experience of Hannibal were outstanding, the Romans won because of their innovative use of different techniques and boarding devices. This battle showed the naval strength of the Carthaginians through the general Hannibal.

Battle of Cape Ecnomus:
This battle was fought in 256 BCE off the coast of Cape Ecnomus in Sicily. It was another notable engagement of Carthaginians in the First Punic War. In this war, the Carthaginian commander, Hamilcar Barca, used his naval strength and expertise to defeat the Roman fleet. This battle had secured a significant victory for the Carthage. In this battle, the Roman admiral Gaius Atilius Regulus and his army used Corvus to drop onto the enemy ships. This battle demonstrated the importance of innovative techniques in ancient warfare.

Battle of Aegates Islands:
The battle was fought in 241 BCE off the western coast of Sicily. This naval battle ended the First Punic War. In this, the Hanno, the Carthaginian admiral, faced the Roman army, which Gaius Lutatius Catulus led. In this battle, the Romans emerged victorious, which led to the peace treaty between Carthage and the Romans, which ended the war. The Carthaginians also showed a good blockage against the Romans. It showcased the importance of naval power and innovative tactics of the Carthage and Romans.

Battle of Lipari Islands:
This battle of Lipari Islands was a naval conflict that happened in 260 BC, in the early stages of the First Punic War. This battle was between the Carthaginian Empire and the Roman Republic. The Carthaginians were under the command of Hannibal Gisco. They defeated the Roman fleet near the Lipari Islands in the Tyrrhenian Sea, north of Sicily. This battle helped the Carthaginians maintain control of various important trade routes. This battle was also a notable victory for the Carthaginians.

Battle of Drepana:
This battle of Drepana occurred in 249 BC, off the coast of Sicily. Hamilcar Barca and Adherbal led the Carthaginians. In this battle, the Carthaginians got a significant victory over the Romans. Publius Claudius Pulcher was there from the Romans's side but was defeated by the Carthaginians. It enhanced the powerful control of the Carthaginians in the western Mediterranean region.

Anecdotes of Naval Warfare and Strategies
Naval warfare has a long history comprising intriguing tactics and innovative strategies during the First Punic War. The Romans faced a big defeat against the Carthaginians in naval warfare. After that, they developed a boarding device that was known as Corvus. It was a hinged bridge with a spike that could easily be dropped onto any enemy ship. This Corvus helped the Roman soldiers to engage in face-to-face combat and board the Carthage Union vessels.

It was a game changer in the battle of Mylae from the Roman side. So, this was a big example of Naval warfare in the past. Also, in ancient naval warfare, ramming was a very common technique. The Greeks used a tactic called diekpal. In this technique, they row through the enemy lines and ram the enemy ships from the sides.

Carthaginian Naval Warfare Strategies

Carthaginians were very diverse and had a significant impact on maritime power.

Boarding Action:

The Carthaginians were well suited for close combat with any other enemy ships. They have employed numerous boarding actions by sending their well-trained sailors to engage in hand-to-hand combat with the enemy ships. In this boarding action, the grappling hooks were used for securing the ships by Carthaginians. It was one of the efficient strategies and tactics employed by the Carthaginians in the past.

Coastal Raids:

Naval forces were involved in coastal raids as well. It included landing marines to attack the enemy coastal cities and territories. It was a common tactic used for territorial expansion by the Carthaginians in their times. It helped them a lot in different battles while destroying the enemy areas.

Use of Mercenaries:

The Carthaginians also used mercenaries during the naval battles. They were skilled fighters who were experienced in Naval fighting. This strategy of Carthage included the hiring of skilled machineries to enhance their naval capabilities and get victory over the enemy crew.

Use of Elephants:

Carthaginians employed numerous strategies in which elephants played a significant role during the Second Punic War. A General named Hannibal from Carthage transported war elephants across the Mediterranean region. It was to reinforce his land campaign against the Roman army. This strategy demonstrates the Carthaginian's creativity and adaptability in using their navy for different purposes.

So, the naval strategies of Carthage contributed to the naval dominance in the Western Mediterranean region. The different anecdotes and strategies offer glimpses into the diverse world of naval warfare. From ancient times to recent history, they all illustrate the importance of adaptability and innovation in sea battles.

Impact of Naval Power on Trade

Naval power has had a very profound impact on trade throughout history. It has helped in establishing and securing the trade routes. The naval forces play a crucial role in ensuring the safety of trade routes by protecting them from various threats. These mainly include pirates. So, their security is vital for the safe transport of the goods and the merchant vessels. The pirates have imposed a very significant threat to the maritime trade. In the past, naval power helped combat piracy by creating a safe environment for trade and shipping.

So, the presence of naval patrols is a deterrent to the pirates. In regions with a higher risk of piracy, the naval forces employed many merchant convoys. This practice ensured that merchant vessels were protected during their journey and that the trade was secured. The nations with strong naval forces expand their trade network by accessing distant markets and establishing new routes. It can lead to economic prosperity and growth.

So, the contemporary and historical significance of Naval power in trade is highly evident for protecting the economic interest in the sea battles. It also fosters economic exchange, facilitates stability, and ensures global trade of goods and resources.

Lessons Applying Naval Tactics in Modern Maritime Security

Applying the Carthaginian naval strategies and tactics to modern Maritime security requires innovation and adaptation. The modern maritime forces can learn different lessons from the Carthaginians, and the strategies can also be adapted for contemporary maritime security.

Embrace Asymmetry and Unconventional Techniques:

The Carthaginians were known for using unconventional tactics like elephants and naval campaigns. It offers many benefits during any battle. So modern maritime security can also benefit from asymmetrical thinking, including cyber capability and innovative technology. It also includes strategies for encountering threats like terrorism, cyber-attacks, and piracy. So, modern maritime trade security should think beyond traditional Naval Warfare to effectively address modern challenges. It will be beneficial for maritime security as well as for the exchange of goods between different regions.

Control of Strategic Areas and Blockade:

The Carthage controlled numerous trade routes through different strategies. These strategies have been helpful for them in safeguarding vital Maritime passages and the trade of different commodities. Similarly, modern Maritime security can also involve the control of chokepoints, including Malacca Strait and Strait of Hormuz. A nation can easily influence regional stability and trade if the modern Maritime companies control these areas. By learning this lesson from the Carthaginians, the modern Maritime forces can easily strategise their plans for controlling the vital maritime trade.

Adaptability and Flexibility:

The Carthaginians were known for their creativity and adaptability in using Naval power. They were very flexible in all their actions of controlling trade. So, in modern Maritime security, adapting to evolving technology and threats is vital. Modern maritime practices should also be flexible. They should be flexible in their naval strategies, including shifting the resources and the techniques of emerging challenges. Adaptability is very important to respond to different threats and challenges.

Combined Land and Naval Forces:

The Carthaginian general Hannibal used naval forces for transporting the elephants and troops. It was a remarkable step that was taken across the Alps. By taking lessons from Carthaginians, modern maritime security can easily benefit from the seamless integration between the land and naval forces. This integration helps crisis management and rapid response in areas prone to natural disasters or conflicts. So, the combined naval and land forces can easily increase trade and enhance maritime security. This strategy should be adopted by modern Maritime security to get benefits and remain safe from the enemies.

Multinational Cooperation and Diplomacy:

The Carthaginians were involved in diplomacy and formed alliances to maintain naval dominance. Similarly, diplomacy and cooperation are the two most important aspects among the nations in modern Maritime security. They should adopt information-sharing collaborative efforts for combating piracy and other threats, and joint naval exercises should be done, which are essential for maintaining maritime stability. The modern maritime staff should be diplomatic and involve different techniques for multinational cooperation.

While modern maritime security differs from naval warfare in ancient times, cooperation, strategic thinking, and adaptability principles are relevant. By recalling lessons from Carthaginians regarding naval tactics, they can easily adopt creative strategies for ensuring global maritime commerce and securing the world's oceans.

Carthaginians were known for their adaptability to naval situations. They were skilled in naval warfare and used many tactics to defeat the enemies. They had developed a unique naval technology which was called Quinquereme. It was very valuable for them in naval battles. Modern maritime security should also adopt new tactics for improving maritime trade. They should combat various threats, including pirates and terrorism, among others. Maritime security should also leverage cutting-edge technology for communication and defence.

CHAPTER 3

CARTHAGINIANS AS TRADERS

The Carthaginians were great traders after being remarkable shipbuilders and supreme naval power. For almost half a millennium, they achieved remarkable success in the bustling trade industry of the Mediterranean Sea. Their achievements were rooted in a well-considered marketing and trading strategy. The important role of Carthage's strategic location cannot be overstated. Nestled in present-day Tunisia, the city of Carthage enjoyed an enviable position at the crossroads of the Mediterranean Sea. This geographical advantage allowed Carthaginians to develop a vast trade network that spanned the Mediterranean region. Their ports served as a crucial transit point for ships carrying goods from various corners of the Mediterranean world, granting Carthage the status of a primary trade hub.

Moreover, their willingness to build strong relationships, uphold ethical business conduct, and adapt to the unique conditions of each region where they traded showcases a comprehensive approach that modern entrepreneurs can learn from. The Carthaginians have paved the way for today's businesses to thrive by demonstrating that building trust, maintaining a reputation, and fostering cultural values are essential for long-term success. Additionally, their dedication to honest and cooperative trade practices is a timeless reminder that fairness, transparency, and ethical conduct are key pillars of commerce. In this chapter, we will explore the marketing strategies of Carthaginian traders that led to their remarkable success in the trading industry. We will also delve into their business ethics and the impact of cultural values on their business.

Marketing and Trade Strategies

For almost more than half a millennium, Carthaginian traders have excelled in the trading industry of the Mediterranean Sea, influencing civilisations and dominating this region with diverse trade. What makes them unique in their business strategy? It's not just luck that showered them with riches, but a well-thought-out trading and marketing strategy that guided them to navigate the waters of the trading industry. While there are no specific documents specifying their marketing and trading strategy, from what we have learned through historical evidence, we can surely discern the marketing and trading strategies Carthaginian traders have benefited from, which also helped them lead a vast trading empire.

Benefit of Strategic Location

One of the greatest advantages the Carthaginians had was their strategic location. Situated in modern-day Tunisia, the city of Carthage held a highly favourable geographical position right at the crossroads of the Mediterranean Sea. This prime location enabled Carthage to establish an extensive trade network covering the Mediterranean region. This strategic benefit played a crucial role in the growth and influence of Carthage in the ancient world, turning it into a central hub for trade and cultural interactions.

Also, ancient Carthage, as one of the colonies of their Phoenician forebears, played a significant role as an import trade port. Carthage was located on a natural harbour that provided safe shelter for passing ships. Consequently, all the ships carrying goods from the western Atlantic coast of Europe and the Mediterranean coasts of Spain, France, western Italy, and western North Africa, en route to the prosperous markets of the eastern Mediterranean and the Middle East, passed through this maritime corridor.

Carthage's control over strategic territories was instrumental in its rise as a prominent trading nation. The possession of vital territories like Sicily, Sardinia, and other Mediterranean islands gave Carthage a firm grip on key trade routes and access to valuable resources. With these territories under their control, Carthaginian traders could safeguard their trade interests and use them as pivotal points for expanding their influence and dominance in the ancient world.

The strategic advantage of holding such territories extended beyond their role as trade hubs. These regions allowed Carthage to exert authority and influence over the surrounding areas on land and at sea. By establishing their presence in these key locations, Carthaginians could secure their interests, protect their trade routes, and exercise control over valuable commodities. This strengthened their position as a bustling port and a leading trading nation, amplifying the impact of their commerce on the broader Mediterranean region.

Diversification of Goods

Although we will delve into the diversity of Carthaginian trade goods and services in the next chapter, we will explore how their diversified trade contributed to their prosperity in the trading empire. It is often said that there was nothing they wouldn't trade for. Some of the most sought-after goods included fine clothes, slaves, and gold jewellery, all considered luxurious and highly in demand during that era.

Another strategic approach employed by Carthaginian traders was acquiring goods from the regions they conquered. For example, they obtained gold from North Africa and sold it at a profitable price to wealthier civilisations.

Moreover, Carthage also recognised the value of products from southern regions of the Sahara, like ostrich feathers, ivory, gold, and black slaves. These items held an exotic allure, making them in high demand within Mediterranean markets. Carthage's strategy was centred on sourcing and trading these exotic goods.

This early investment in luxurious items allowed them to generate significant profits and helped them establish a monopoly in the industry.

Improve the Naval Power

As we extensively covered Carthage's naval power in the second chapter, we'll explore how this naval strength became an integral part of their trading strategy. Carthage faced continuous challenges from the native inhabitants of the regions they had settled, particularly the Berbers. Many of these Berbers would later establish influential kingdoms like Numidia and Mauritania. They eventually formed alliances with Rome to overthrow Carthage, which fell under Roman rule.

Consequently, from the beginning, Carthage had to establish and maintain a strong military presence, which later evolved into one of their greatest assets. Additionally, Carthage's formidable naval fleet was crucial in safeguarding vital trade routes and countering rival powers. For instance, they controlled important waystations in southern Spain, including Sicily and Gades (Cadiz). In this way, Carthage turned its initial vulnerability into a formidable strength, ultimately enabling it to protect and empower its trade networks.

Colonial and Trade Routes Expansion

The Carthaginians were not just great traders; they embodied the spirit of warriors and explorers. They had a fearless willingness to explore new trade frontiers and take calculated risks, which led them to discover fresh trade routes. For instance, Carthage established colonies in new territories like Sicily and the Iberian Peninsula. These colonies served the dual purpose of protecting their trade interests and maintaining control over these markets, thus expanding their influence.

Moreover, Carthage ventured into the Iberian Peninsula to pursue silver and tin. These metals were essential for producing bronze, a highly important material in that era. They conducted trade overland across the Sahara and by sea to the Cassiterides to obtain tin. The challenge was that their tin sources, located in the Canary Islands and the British Isles, were distant, and the Atlantic Ocean was largely uncharted.

Additionally, Carthage's explorations and voyages along the west coast of Africa highlight their strong desire to access exotic goods from these regions and establish them as their trading territories. This adventurous and enterprising spirit contributed significantly to their success as traders and explorers.

Building Strong Trade Relationships

Building a strong and trusting relationship with Carthaginians with their partners highlights the greatness of Carthaginian traders. They were known for maintaining high standards when bartering their goods and were patient until they were content with the price they received for their products. What's even more impressive is that they conducted these transactions without any conflicts or disputes. A great example of this is the Greek historian Herodotus, who detailed the practice of trade rituals.

"The Carthaginians unlade their wares, and having disposed of them after an orderly fashion along the beach, leave them and return to their ships, raise a great smoke, come down to the shore, and, laying out to view so much gold as they think the worth of the wares, withdraw to a distance.

The Carthaginians, upon this, come ashore and look. If they think the gold is enough, they take it and go their way. But if it is insufficient, they go aboard the ship again and wait patiently. Then the others approach and add to their gold till the Carthaginians are content."

This incident shows the excellent relationships Carthaginians had with other traders. It demonstrates the trust others placed in them, as they were willing to exchange their goods for gold, an expensive and valuable metal throughout history.

Furthermore, Carthaginians were skilled at maintaining trade treaties with other nations to build strong and lasting relationships. For instance, they had agreements with Rome at various times, like in 509 BCE and 348 BCE, to establish exclusive trade zones. These agreements allowed them to have control over trade in those regions. Such trusting relationships and treaties with their trading partners were instrumental in helping Carthaginian traders maintain long-term businesses and achieve great profits.

Seafarerers and Shipbuilding skills as a Trading strategy

As discussed in the first chapter, the Carthaginians excelled in shipbuilding and transformed traditional Phoenician ships into warships and cargo vessels. The purposes of shipbuilding varied for the Carthaginians, but one thing was certain: they understood the immense impact of shipbuilding on trade.

Advancements in shipbuilding techniques allowed them to construct larger and more efficient vessels. These innovations enabled Carthaginian ships, like the large hippos and gaulos, to be tailored for maritime trade. Furthermore, this enhancement in ship design improved their capacity to transport goods efficiently. As a result, they managed to reduce the cost per unit and enhance their overall profits.

Not only this, their exceptional seafaring skills also played a crucial role in adapting ships with long-range capabilities like Carrivals. These advancements allowed them to take risks by venturing into the deep waters and exploring uncharted lands. This adventurous spirit enabled them to establish trade routes in distant regions and expand their geographical scope.

Overall, without their remarkable seafaring and shipbuilding skills, the Carthaginians could not have built the trading empire that profoundly impacted the trade history. Their willingness to explore new horizons and navigate challenging waters was key to their success.

Adaptation to Regional Conditions

The Carthaginians, known for their sharp wits and entrepreneurial mindset, recognised the importance of adaptability to different regions. When acquiring goods, they didn't impose their ways but understood the value of following local practices. This approach not only earned the trust of local traders but also made their trading activities more convenient.

For instance, large-scale trade across the Sahara Desert was initially impractical due to the absence of camels in the region. However, by adapting to the local conditions, they found a way to conduct trade through local exchanges. This strategy allowed for steady and successful trade growth in the region, ultimately enabling them to obtain precious items like ostrich feathers, ivory, gold, and black slaves.

Using their home Resources

Carthage possessed not only a strategic location as a trading port and colony but also significant geographical and production advantages that boosted profitability. Furthermore, various industries enabled them to produce high-quality goods sold at a premium in other regions.

For example, archaeologists have discovered evidence that Carthaginian traders further developed the purple dye industry, a signature product of their Phoenician forebears. Purple dye was highly valued and expensive during that era, which contributed to their increased profits.

In addition to purple dye, the Carthaginians also excelled in producing high-quality embroidered clothing for export. Furthermore, they strategically decided to cultivate crops such as wheat, olives, grapes for wine, dates, figs, pears, and pomegranates on their land. They practised careful irrigation and animal husbandry to ensure the health of their crops and livestock. These agricultural products were then sold in ports all around the Mediterranean Sea.

Overall, Carthaginian traders demonstrated their worth through their exceptional trading and marketing strategies, leaving valuable lessons for modern traders to learn from their high-value approach to commerce.

Ethical Trade Practices and Fairness

Simply put, ethical practices served as a key marketing strategy for the success of Carthaginian traders. However, in this category, we will delve into the business ethics of Carthaginian traders, which played a pivotal role in building trustful relationships with their trading partners. Their commitment to ethical conduct was a significant factor in their success in trading.

Fair Dealing and Honesty in Trade

The Carthaginians were known for their trade fairness and commitment to equal treatment for their trading partners. A prime example of this aspect is how they treated foreign traders in their homeland. Given their strategic location and bustling ports that attracted ships from various other traders, the Carthaginians extended invitations to traders from other nations for open trade in their land. This hospitality was extended to traders from places like Rhodes, Athens, and Italy.

Most importantly, they treated these foreign traders with the same respect and fairness as their city's merchants. This practice underscored their commitment to fairness and equality in trade, making them trustworthy partners in the trading world.

The Carthaginian Code of Ethics in Commerce

The writings of Herodotus shed light on the Carthaginian barter practice with native traders along the North African coast, which served as a model of fairness. For instance, the Carthaginians would welcome native traders with a smoky signal, displaying their goods for all to see. Native inhabitants would approach, offering gold in exchange for the goods.

It is important to note the Carthaginians' steadfastness in determining the price and value of their goods. This practice was rooted in their strong business ethics, as they knew the value of their goods and stood firm in demanding the price without compromise. They only accepted gold when it equalled the value of their goods, and the natives refrained from touching the goods until the gold was paid. This ethical approach emphasised equitable and transparent trade, benefiting both parties. Their unwavering pricing values also helped them make more profit and communicate their trading terms clearly to other traders.

Trusting Relationships Through Trade Treaties

Carthaginian traders didn't just form alliances but also engaged in making treaties and diplomatic relationships with many nations. While treaties were significant, what truly stood out was their willingness to establish agreements (treaties) with other nations and actively participate in bustling markets in cities like Athens, Delos, and Syracuse. This demonstrated their genuine commitment to fairness and collaboration with other influential trading partners.

Their dedication to honest and cooperative trade practices played a crucial role in building strong relationships, ultimately leading to their success in the world of trade. Their approach was a testament to their reliability and ethical conduct in the business world.

Cultural and Regional Influence on Trade

The Impact of Religion on Carthage's Trade

The Carthaginian religion, derived from Phoenician ancestors, was based on polytheism. Interestingly, it has a great influence on the trade as well. For example, this religious diversity added a unique layer to Carthaginian culture and directly impacted trade. It was not merely a matter of faith; it significantly influenced their commercial decisions.

In simpler terms, the Carthaginian religion was a diverse tapestry of deities, many localised and known only under their regional names. Therefore, according to their religion, certain deities were associated with specific regions, affecting the choice of trading partners.

Influence of Carthaginian's culture on Trade

Much of Carthaginian production centred around artistic embroidery in clothing and pottery. Similarly, Carthaginian artisans excelled in working with various materials, including ivory, glassware, wood, metals, and precious stones. The high-quality artistic style influenced by their culture made their goods highly desirable in the global market, providing a solid foundation for their trading partnerships.

The Fusion of Cultures along Exploration

It is important to note that embroidery is not the sole component of Carthaginian culture. Their exploration and adventurous spirit are also integral aspects of their cultural inclinations, which later played a crucial role in establishing them as the central maritime power in the Mediterranean. Furthermore, their trading empire extended across the Mediterranean, perhaps even into the Atlantic, highlighting their cultural affinity for the sea. For instance, trading goods like gold, ivory, and other luxurious items from the Sahara Desert held a special cultural significance for Carthaginian traders.

Lessons for Modern Businesses in Ethical Trade and Marketing

With their innovative marketing strategy and intriguing religious influence on trade, the Carthaginian traders conveyed a profound message to their predecessors and the modern world. Despite being a relatively small civilisation, they demonstrate that one can exert dominance in trade across continents and beyond through astute wit, well-planned strategies, and upholding deeply ingrained cultural values. Consequently, these are the vital lessons bequeathed by the Carthaginian traders:

Developing Persuasive Sales Strategies

A persuasive and effective sales strategy is the best friend of entrepreneurs, as it not only helps businesses understand their customers' desires but also lays the foundation for future effective business plans. Carthaginian traders exemplified this approach; they sold goods where traders knew the true value of their goods, thereby saving valuable time. Additionally, this approach motivated them to channel their efforts into producing high-value goods. Modern businesses can adopt a similar strategy by comprehending their customers' mindset, ultimately increasing sales and success.

Strong Approaches in Marketing

The Carthaginians were known for spreading tales of exploration and voyages through storytelling, a remarkably effective technique for attracting more customers. Moreover, the high quality of their goods served as another solid foundation for marketing their products. For instance, people from other civilisations were so eager to trade with them that they valued their goods as highly as precious metals like gold. To achieve such a potent marketing approach, the Carthaginians provided goods of exceptional quality, earning them a strong reputation and goodwill. In the modern business world, companies can invest in products that tell the stories of their quality rather than relying solely on random marketing channels.

Importance of Ethical Trade

We have already explored the ethical and fair trade practices of the Carthaginians. We also know that their success in building a vast trading empire was not solely attributed to their hard work and strategies but also their ethical business approach. This foundation of implanting business ethics enabled them to foster strong relationships with their trade partners, resulting in seamless trade in the future. Modern businesses can undoubtedly adopt this firm and steadfast business ethics to uphold their values and maintain lasting relationships with their trading partners.

Building Trust and Reputation

The Carthaginians' commitment to business ethics was more than just a footnote in their success; it was a cornerstone. Their ethical practices were not limited to their interactions with trading partners but extended to every aspect of their trade. This ethical approach earned them the trust of not only their counterparts in the business world but also their people, boosting profit and monopoly of the industry effortlessly.

Modern businesses can also learn the importance of such a reputation from Carthaginian traders. Companies that uphold ethical standards treat their employees fairly, and engage in ethical trade practices often find that their goodwill becomes a valuable asset. Customers increasingly value businesses that demonstrate a commitment to ethical conduct. This trust, once earned, can lead to long-term success and a loyal customer base.

Recognising the Cultural Impact of Trade

Culture is more than just a name of social and ethnic norms; it can be built upon good habits. For traders, it's about creating a culture that helps navigate the complex waters of the business world. In this context, Carthaginian traders were fortunate to have a cultural foundation rooted in maritime traditions and aesthetic sensibilities.

Modern businesses can draw inspiration from this by actively promoting cultural values within their organisations. By fostering a culture that values ethics, innovation, and excellence, companies can strengthen relationships with their employees and customers. This leads to growth and success and creates a positive work environment that benefits the organisation and its employees. Building a culture that aligns with the goals and values of the business is a powerful strategy for modern companies.

Overall, these ancient traders from Carthage provide us with a wealth of insights into building thriving businesses. Their enduring success was not a stroke of luck; it resulted from well-thought-out strategies, a strong commitment to ethical practices, and a deep understanding of the importance of cultural values. They were masters of persuasion in their sales strategies, masters of marketing, and staunch defenders of fairness in trade. Carthaginian traders stand as a testament to the power of innovation, astute strategies, and an unwavering commitment to ethical trade, providing modern entrepreneurs with a valuable blueprint for success.

Chapter 4

SHIPPING AND TRADING GOODS

The Carthaginians, known for their exceptional trading skills, had a remarkable reputation for dealing with almost anything. Their willingness to sell to anyone who had the means set them apart. They primarily focused on trading exotic and valuable goods in high demand. Despite the diversity of goods they handled, their mercantile expertise was so extraordinary that traders were willing to pay their prices without inspecting the goods. This showcases their strategy's effectiveness and commitment to delivering high-quality products, ultimately leading them to become the wealthiest civilisation of their time. In this chapter, we'll delve into the kinds of goods the Carthaginians traded and their remarkable success through their strategies with various merchandise.

Notable Carthaginian Trade Goods

The Carthaginians traded many, including precious metals, foodstuffs, slaves, and high-quality manufactured items such as exquisite textiles and gold jewellery. Many of these items were exotic and highly sought after, ensuring significant profits for the Carthaginians. Here is a list of the goods that Carthaginian traders were involved with:

Precious Metals and Monopoly of Tin

Like most prominent civilisations of that time, the main source of trading for Carthaginians was acquiring precious metals. However, along with trading gold and silver, other metals like tin, copper, lead, and iron were also treasured due to their various uses in several industries and products. Carthaginians were especially interested in tin, and they monopolised the tin industry. Tin was super important because it was needed to make bronze, which was highly valuable in those days. However, acquiring tin was a big challenge because Carthage had to obtain it from faraway places, such as the Canary Islands and the British Isles. The problem was that, back then, nobody had a map of the Atlantic Ocean, so it was quite an adventurous risk that Carthaginians had to take to acquire the tin. In this way, tin, along with silver, was traded to other Phoenician colonies.

The Carthaginian traders employed a brilliant marketing strategy regarding precious metals. They acquired new territories in search of precious metals like gold and silver and traded them for the locals' desired goods. By doing this, they could sell their goods at higher prices. Carthaginian traders also exported exotic art pieces crafted with silver and gold from their industry. Their wealth is evident from the large armies of hired soldiers they could support. This approach boosted their wealth and helped them maintain valuable trade relationships across different regions. In fact, during the 4th century, Carthage minted a substantial amount of gold coins, much more than what was typical for other advanced nations. Besides precious metals, Carthaginians also engaged in trade for precious stones, with carbuncle being a privilege to become a sought-after gemstone among the Carthaginians due to its abundance in trade. This further emphasised their involvement in trading precious metals and establishing some degree of monopoly in these industries.

Spices and Foodstuff

Carthaginians engaged in the trade of spices, recognising their value and importance across different cultures. Spices were sourced from regions such as the Arab world, India, and Africa, and Carthaginian traders played a vital role in bringing these valuable commodities to various markets. In addition to spices, salt was a crucial commodity, especially for inland tribes. Carthaginians incorporated salt into their trade routes to cater to these tribes' needs, demonstrating their adaptability and willingness to meet the demands of various customer bases.

Furthermore, Carthaginian traders dealt with everyday essentials, including olives, a dietary staple, and their precious oil, indicating a diversified trade portfolio encompassing luxury and staple goods. This approach allowed them to cater to various consumers and maintain a successful trade network.

The aroma of Carthaginian wine, highly regarded in the ancient world, was widely savoured. Cereals, the breadbasket of many ancient civilisations, were also among their exports. Additionally, they brought forth the flavours of salted fish, aromatic garlic, sweet pomegranates, nourishing nuts, fragrant herbs, and a medley of exotic spices that enriched the Mediterranean trade network.

Archaeological findings provide fascinating insights into the Carthaginian trade, exemplified by two Punic shipwrecks. The discoveries shed light on the ancient spice and foodstuff trade, highlighting its significance in connecting different regions and cultures.

One shipwreck, dating back to the 5th century BCE near Ibiza, contained a precious cargo of fish sauce known as "garum." This delicacy would later become a favourite of the Romans, underscoring the influence of trade on culinary preferences.

The second shipwreck, located off the coast of Marsala in Sicily and dating to the 3rd century BCE, carried not only amphorae of wine but also a bounty of olives. These findings emphasise the interconnected world of ancient trade, revealing that even humble items like fish sauce, wine, and olives played a significant role in the maritime exchange, enriching our understanding of the past.

Textiles and Dyes

Carthage was a prominent production hub for fine textiles and was renowned for its embroidered fabrics. The garments crafted in Carthage held exceptional value in the market thanks to their intricate and artistic embroidery. Skilled artisans from the Carthaginian industry dedicated their expertise to producing these exquisite textiles, making Carthaginian clothing highly sought after for its craftsmanship and artistry.

The Carthaginian traders continued the tradition of their Phoenician ancestors by engaging in the production and trade of a special purple dye made from shellfish. This dye, which originated from the Phoenician colony of Mogador on the northwest coast of Africa, was highly esteemed and in great demand in the markets of the ancient world. Carthaginian traders contributed to the popularity and circulation of this valuable purple dye across various regions.

Ivory and Exotic Animals

Carthaginian traders were crucial in exporting and importing commodities through Carthage's port. One of the significant items they traded was ivory, which was important due to its utilisation in producing a wide range of goods, such as jewellery and art pieces. The import and export of ivory were central to Carthage's economic activities, contributing to its prosperity and influence in the ancient world.

The origin of ivory used by the Carthaginians is partially clear, but they likely engaged in extensive trade of this precious material during their prosperous times. Africa was known for having some of the largest and finest ivory, and African elephants, which provide ivory, were present throughout the continent. However, elephants have been pushed away in areas where human presence encroached. Interestingly, the Carthaginians even managed to domesticate elephants, an achievement not matched by any other African group. This allowed them to acquire high-quality ivory for industrial and trading purposes, highlighting their resourcefulness in the ancient world.

Other Merchandise

Slavery was a grim reality in ancient times, and Carthage was no exception. The practice involved forcing individuals from various backgrounds, including those from respectable families with education, into servitude, often due to being prisoners of war or conquering their towns. Carthaginians recognised the importance of having obedient and docile slaves, treating them well to ensure their cooperation and productivity. This allowed them to trade in human bondage successfully.

In addition to slaves, Carthage engaged in trade with a wide array of goods. This included weapons, tools, cooking and eating utensils, scissors, amulets, jewellery, exquisitely designed glassware, wooden furniture, ceramic figurines, decorative ostrich eggs, incense burners, and ornamental masks. The diversity of goods highlights the expansive and varied trading practices of the Carthaginians, who sought to acquire a wide range of items from different regions.

Overall, the extensive list of Carthaginian products and trading goods underscores their successful engagement with various commodities. Carthage's ability to forge trading relationships and import and export goods from various regions speaks to its prominent role in the ancient trading world. The Carthaginians' influence and impact were felt through their craftsmanship and ability to connect with different cultures and markets. This diverse and interconnected trade network was vital to Carthage's prosperity and economic influence in the ancient world.

Lessons in SpecialiSation and High-Value Trading

These are the remarkable lessons Carthaginian traders left for modern businesses:

SpecialiSation and Expertise

One of the major strengths of Carthaginian traders was their ability to specialise in specific types of goods while still dealing with a wide variety of products. In other words, although they traded in many goods, they had particular expertise in certain categories, such as tin and purple dye. This expertise allowed the Carthaginians to have an in-depth knowledge of these goods and sell them to traders who recognised their value. Furthermore, by becoming specialists in these particular products, they earned a strong reputation in the market for these goods.

Modern businesses can draw valuable lessons from this approach and consider becoming experts in specific products before diversifying into others. This specialisation can help them stand out in the market and build a reputation for quality and knowledge in their chosen field.

Carthaginian Craftsmanship

History has shown that the Carthaginians greatly appreciated the arts and excelled in producing embroidered clothing and artistic pieces on an industrial scale. This means that for thousands of years, there were textile and art industries with skilled artisans who meticulously crafted these fine works of art by hand. Moreover, there was a high demand for these pieces in the global market, highlighting how much other civilisations valued their craftsmanship.

Modern businesses can learn from this aspect by infusing aesthetics into their products. By focusing on high-quality craftsmanship and appealing to customers who appreciate such fine work, businesses can attract a dedicated customer base willing to pay a premium for quality and artistry. This approach can help modern businesses stand out and create a strong market presence.

Diversity of Goods

Dealing with various types of goods in the market can be risky, especially for new businesses. However, numerous modern examples of businesses successfully offer various products. Amazon and Alibaba are excellent contemporary examples. In the ancient world, the Carthaginians were a prime example of this diversity in trade. They traded many goods, from salt to purple dye, and were nearly unstoppable in their trading endeavours.

What's particularly intriguing is that the Carthaginians excelled in each product. They knew where to source specific items and where to find the right markets for them. This strategic approach was a key factor in their success.

Modern businesses can learn from this strategy and use it as a roadmap for their success. If they aim to deal with a diverse range of goods, they can follow in the footsteps of the Carthaginians to become successful traders across various aspects of business. This strategic approach can guide them in navigating the complexities of the market and achieving success with various products.

High-Quality traded Goods

Quality is a critical factor in the success of any business, regardless of how diverse your product offerings are or how much expertise you have in particular goods. If your product quality doesn't meet the mark, all your other strategies may become ineffective, as low-quality goods often have little to no profit. The Carthaginians understood this principle thousands of years ago and made a significant effort to provide high-quality goods to their trading partners. This commitment to quality was one of the reasons their trading partners had complete trust in their dealings and valuable items.

Modern businesses can learn from the Carthaginians by placing a strong emphasis on the quality of their products. By doing so, they can build a trusting relationship with their business partners, just as the Carthaginians did. This not only ensures that they can meet the needs and expectations of their customers but also helps them establish a reputation for reliability and quality in the marketplace.

In conclusion, it's highly unlikely for a civilisation to become the wealthiest of its time without employing strategies and sharp wits. Carthaginian traders handled diverse goods, leveraging their expertise in specific categories. They devised remarkable trading strategies, securing a virtual monopoly in the metal industry, particularly regarding tin. Additionally, the Carthaginians skillfully navigated the complexities of the market, trading a wide range of items, from precious metals to spices and foodstuffs, excelling in every product they engaged with. Their success resulted from a combination of astute planning and deep industry knowledge.

Chapter 5

TRADE PARTNERS AND ALLIANCES

Trade and diplomacy formed the foundation of the influential Carthaginian civilisation, which became prominent in the Mediterranean region for nearly 600 years. By actively cultivating an intricate web of strategic economic and geopolitical partnerships across the known world, Carthage established itself as a preeminent maritime trading powerhouse. In the bustling markets of the historical world, the Carthaginians were renowned masters of trade. They navigated the Mediterranean Seas, organising a big community of exchange companions and alliances that extended from their North African stronghold throughout the waters to Europe, Asia, and beyond. Their ability to forge beneficial relationships with neighbouring tribes and distant civilisations enabled them to secure precious assets and particular merchandise, enriching their lifestyle and expanding their influence.

This chapter will thoroughly examine the extent of Carthaginian trade routes, goods exchanged, and nuanced approaches to alliance-building. It will provide an in-depth analysis of the multifaceted dynamics of Carthaginian treaty-making and diplomacy that successfully cultivated durable, mutually beneficial relationships with diverse foreign peoples and states. We will also examine the lessons contemporary agencies can learn from their trade partnerships and alliances.

Carthage's trade partnerships and networks

Carthaginian traders established successful commercial alliances with major foreign partners through proactive diplomacy and treaty-making. A pivotal 6th-century BCE treaty with the Etruscans granted Carthaginian merchants extensive access to Etruscan ports and cities in exchange for concessions in Sardinia and Corsica. This allowed Carthage to tap into vital Italian trade networks efficiently. Alliances with Greek Sicilian cities like Syracuse gave Carthage agricultural goods and strategic Mediterranean ports. Treaties with tribes in Iberia, North Africa, and Sardinia opened new sources of metals, livestock, and other commodities— even alliances with rivals, like Rome, aligned spheres of influence to enable unfettered Carthaginian trade. By integrating diverse groups into its trading ecosystem via win-win partnerships enforced by formal accords, Carthage created the dominant commercial power in the Western Mediterranean for centuries. Its traders leveraged bonds of mutual interest to collaborate with former competitors and unfamiliar cultures successfully.

Founded circa 814 BCE on the North African coast in modern-day Tunisia, Carthage rapidly grew into a major maritime trading hub, given its strategic location on key Mediterranean trade routes. Its merchant fleets crisscrossed the Mediterranean Sea, establishing far-reaching trade ties with city-states across Greece, Italy, France, Spain, and North Africa. Core trading partners included major Phoenician cities like Tyre, Sidon, and Cadiz that shared cultural bonds with

Carthage. Additional Greek cities such as Syracuse, Athens, and Corinth, and emerging coastal Italian cities like Rome, also became integral trade partners.

In the Western Mediterranean, Carthage conducted lively trade with the numerous tribal groups inhabiting the Iberian peninsula, North Africa, Sicily, and Sardinia. Carthage leveraged maritime trade and overland caravan routes across the Sahara to exchange goods with interior African kingdoms and tribes. These groups provided commodities like gold, ivory, and exotic spices.

Carthage deployed common business practices like standardised weights and measures, currency valuations, accounting methods, and legal codes to manage trade across its vast networks efficiently. This enabled smooth transactions with diverse partners accustomed to Carthage's commercial conventions. Warehouse receipts, bills of lading, and other documentation facilitated trade even with remote partners divided by language and cultural barriers.

Carthage also proactively disseminated geographical and navigational knowledge to expand its trade horizons. It sponsored bold voyages of maritime exploration along the Atlantic coast of Africa and Europe. Carthaginian merchants eagerly sought new sources of commodities and potential customers. This appetite for pushing boundaries allowed Carthage to integrate previously isolated groups into its sprawling network. Generally, Carthage's proactive efforts to integrate disparate groups into its trading ecosystem provided major competitive advantages.

Carthage's Diplomacy and Treaty-making

Carthage placed immense emphasis on proactive diplomacy and formal treaty-making to cultivate enduring political and economic alliances with major foreign powers. It regularly dispatched diplomats and envoys to nurture relationships with foreign rulers, governments, and elite groups. These experienced envoys negotiated elaborate quid pro quo pacts that granted Carthaginian traders extensive legal rights and protections when conducting business abroad.

For example, in the 6th century BCE, Carthage signed a commercial treaty with Italy's Etruscan League of City-states. This treaty offered specific territorial concessions in Sardinia and Corsica to the Etruscans in return for allowing unfettered Carthaginian traders access to vital Etruscan coastal cities and ports. Carthage also signed formal treaties with the emerging Roman Republic to clearly define mutual spheres of military influence and control and prevent unnecessary conflicts. Even when dealing with less powerful groups like African and Iberian tribes, Carthage preferred using astute diplomacy and deal-making rather than brute force alone to secure voluntary trade partnerships.

The Carthaginian ruling council and executive magistrates astutely recognised that diplomacy and treaty-making were far more cost-effective strategies for expanding trade and political influence than military conquest alone. While Carthage did build a formidable navy and army to protect far-flung trade routes and colonies, prolonged wars and military occupations were ruinously expensive and ultimately counterproductive for maximising profits. Savvy Carthaginian leaders realised it was more prudent to use diplomacy, incentives, and formal pacts to elicit voluntary cooperation from foreign powers whenever feasible.

Carthage deployed three main types of envoys to carry out this proactive diplomacy and treaty-making. The first were special representatives appointed directly by the Carthaginian Senate or executive Suffets to handle high-stakes negotiations with powerful foreign rulers and governments. These plenipotentiary envoys had the authority to hammer out comprehensive political and commercial treaties on Carthage's behalf without constantly sending them back for official approval.

The second type was magistrates elected to serve one-year terms as roving commercial envoys promoting Carthaginian trade interests abroad. These lower-level diplomats focused on building goodwill and securing concrete privileges for Carthaginian traders operating within foreign seaports and commercial centres. The third type was multilingual local messengers and intermediaries recruited from Carthage's widespread ports and colonies. These agents set up introductions with foreign leaders and helped smooth negotiations to finalise accords.

The extensive diplomatic outreach pursued by these Carthaginian envoys served several key strategic purposes. First, it enabled Carthage to actively monitor political and economic shifts within foreign societies that might impact its overseas trade and colonies. Second, it allowed Carthage to constantly lobby foreign governments for new concessions to boost Carthaginian merchant activity and profits further.

Third, regular diplomatic contact nurtured personal relationships between Carthaginian envoys and foreign leaders. This fostered greater mutual understanding and goodwill, making both sides more inclined to resolve disputes through negotiation rather than conflict. Fourth, signing formal written treaties codified the rights of Carthaginian traders operating abroad, giving them stronger legal recourse against unfair exploitation or commercial obstruction.

Carthage's extensive diplomatic outreach and treaty-making resulted in a vast network of bilateral and multilateral commercial accords with key partners by the 4th century BCE. Major examples included treaties with the Etruscan League, Greek Sicilian colonies, Sardinian tribes, and the Roman Republic. Each treaty contained specific clauses outlining trade concessions, territorial boundaries, military aid obligations, arbitration procedures, and other details to place commercial and political relations on an equitable, contractual footing.

Overall, Carthage's energetic diplomacy and negotiation of formal international agreements reflected the city-state's fundamental commercial outlook and priorities. Regular diplomatic contact and detailed written treaties minimised trade disputes and other disruptions to commerce, promoting maximum profit. Yet Carthage's dependence on skilful diplomacy proved a longer-term liability once its military power waned. When diplomacy failed, Carthage often lacked adequate fallback options to defend its remaining trade empire and colonies forcibly.

Noteworthy Case Studies of Alliances
Carthage's robust partnership with ancient Egypt represents an outstanding model of symbiotic alliance. Lasting nearly 300 years, from 664 to 373 BCE, it bolstered civilisations immensely through vital resource exchanges. Egypt helped supply massive grain surpluses to help feed Carthage's populous cities. In return, Carthage provided Egypt access to Phoenician manufactured goods and reinforced Egypt's navy with its maritime forces. Carthaginian diplomats cemented these ties by intermarrying with the Egyptian elite. Despite internal upheavals, this alliance endured due to its fundamental mutual benefits.

Carthage also shared deep-rooted cultural, religious, and linguistic bonds with the scattered Phoenician city-states across the Mediterranean coasts. This facilitated generally cordial trade ties from Carthage's earliest days. Carthage relied heavily on timber, textiles, and glass manufacturing know-how imports from older Phoenician cities like Tyre, Byblos, and Sidon. Joint colonies like Ibiza, Cadiz, and Utica were established to extend both groups' reach. The Carthaginians eagerly adopted and enhanced Phoenician innovations in shipbuilding, sails, and navigation to eventually dominate maritime trade.

Carthage successfully forged profitable commercial pacts in the Western Mediterranean with Iberian, North African, Sicilian, and Sardinian tribal groups through adroit diplomacy. These accords provided Carthage with ample precious metals, minerals, food, and soldiers in exchange for advanced manufacturing, luxury goods, and weapons. For instance, Iberian alliances granted direct access to rich Spanish silver mines, enabling much of Carthage's trade and naval activities. Even former foes like Syracuse were eventually won over through persistent diplomatic efforts after 480 BCE.

Lessons for Modern Businesses Building strong interpersonal ties
Carthage's success provides many timeless lessons for contemporary businesses seeking to establish viable strategic partnerships. As Carthage cultivated personal relationships between its envoys and foreign leaders, investing efforts to foster strong interpersonal ties and affinity with partners remains vital. Seconding employees to partner organisations facilitates building relationships at the ground level. Providing reciprocal staff training, technical support, and preferential pricing builds durable goodwill between partner teams.

Framing Partnerships as Voluntary Win-Win Associations
Carthage framed its partnerships as voluntary associations based on mutual benefit, incentivising win-win collaboration with allies. Similarly, modern firms should position their alliances not as exploitative or coercive but as opportunities for joint value creation through voluntary participation and carefully aligned incentives.

Developing formal contracts to align responsibilities
Carthage negotiated formal written treaties to codify the rights and responsibilities of each party in an alliance. This prevented future misunderstandings. Likewise, modern businesses should develop contracts, MOUs, and agreements to provide a clear framework for aligning responsibilities within a partnership.

Leveraging Geographic and Cultural Proximity
Carthage's alliances emphasised partners with geographic and cultural proximity in the Mediterranean region, which eased interactions. When feasible, today's companies should also consider allying with accessible partners in familiar regions, leveraging geographical and cultural common ground.

Pursuing alliances with competitors
Carthage was pragmatic in allying with rivals when it served mutual interests. Similarly, startups can ally with incumbents for market entry, while even big competitors pursue partnerships to co-market and jointly grow demand. Alliances need not be limited to non-competitors.

Leveraging Alliance Networks
Carthage created an expansive network of bilateral and multilateral alliances. Today's firms should similarly cultivate a diverse alliance portfolio with suppliers, distributors, competitors, and complementors. This provides resilience and flexibility to adapt to changing conditions.

Treating alliances as dynamic relationships
Carthage constantly monitored its alliances and dispatched envoys to renegotiate terms when needed. Companies today should also consider their partnerships dynamic, requiring ongoing communication, adaptation, and recalibration as market contexts shift over time.

Maintaining control and sovereignty

Even while allying, Carthage remained fiercely protective of its independence and sovereignty. Similarly, modern firms should be careful not to over-integrate or cede too much control to partners, maintaining flexibility to pivot strategies.

Focusing on relevant competencies

Carthage focused diplomatic efforts on partners that could provide meaningful capabilities lacking internally. Companies should similarly evaluate gaps in competencies and resources and seek alliances that fill those specific needs.

Avoiding Overdependence on Alliances

While beneficial, Carthage's reliance on alliances proved a liability once its power waned. Firms today should avoid overdependence on partnerships and maintain alternative options.

Summarising this chapter, Carthage represents an instructive historical case study of how an entire civilisation heavily relied on actively creating strategic international trade partnerships and alliances to sustain its prosperity for centuries. Diplomacy and deal-making were as essential as military might to Carthage's dominance of Mediterranean trade networks. Alliances with Egypt, Phoenicia, Iberian, and North African groups stand out as models of mutually beneficial relationships built on aligning incentives for cooperation. Although Carthage ultimately faded away, its legacy endures through timeless principles modern businesses can emulate to build successful, durable, globally competitive strategic business alliances. The Carthaginian experience remains an illuminating example of the tremendous value created when enterprises choose cooperation through visionary relationship-building over conflict.

Chapter 6

TRADE ROUTES AND TRADE PORTS

Strategically situated on Africa's northern shores, Carthage dominated maritime trade networks, enabling the exchange of treasures from as far as the fabled tin mines of Britain to the exotic riches of sub-Saharan Africa. Carthage's formidable warships protected trade routes from piracy while their colonies fed raw materials into a thriving mercantile economy. Follow the intrepid Carthaginian sailors as they navigate stormy seas, fend off pirates, and venture beyond the Pillars of Hercules into the unknown. Their voyages along expansive trade routes fuelled an empire's prosperity and shaped history in the ancient Mediterranean and beyond.

In this chapter, we'll delve into the trade routes and colonies explored by Carthaginians, their impacts on trade and their legacy. Additionally, we'll look at the lessons modern businesses can learn from Carthaginians' explorations.

Carthaginian trade routes and their impacts on trade

The Carthaginians established an extensive trading network across the Mediterranean Sea, instrumental to their rise as one of the wealthiest civilisations in antiquity. They built a network of strategically located ports that served as major commercial hubs connecting trade routes across their sphere of influence. The two most vital ports were at Carthage along the North African coast and Cadiz in southern Spain. These ports provided access to Carthaginian territories and colonies scattered along the coasts of North Africa, Spain, Sardinia, Sicily, and the Balearic Islands.

Goods from across the known ancient world flowed into Carthage's bustling ports. Precious metals like silver, tin, and iron were imported from as far as Britain and West Africa to supply Carthaginian coinage and tools. Fine Egyptian linens and papyri were prised for sails, clothing and record keeping. Olive oil, wine and artisanal goods came from Greece, Italy and Carthaginian territories. Arabia and East Africa provided spices, incense, ivory and exotic animals. In turn, the Carthaginians exported locally made textiles, pottery, glasswork, and agricultural products from their colonies.

Their trade routes stretched along the Atlantic coast up to Brittany and Britain, while other routes connected various coastal cities in North Africa. Carthage maintained extensive maritime trade with Phoenician colonies across the western Mediterranean, exchanging raw materials, manufactured goods, foodstuffs, and new ideas. Their powerful navy protected the safe passage of merchant ships, which kept piracy in check. The Carthaginians reaped enormous wealth by controlling and taxing the flow of goods through their ports. Their trading economy funded the growth of Carthage into a dominant commercial and military power for centuries in the ancient Mediterranean world until the Punic Wars with Rome led to their eventual decline. The Carthaginians' vast maritime trade network was the lifeblood that sustained their civilisation, wealth, and imperial ambitions.

Main Trade Routes That Fueled Carthaginian Prosperity

The Carthaginians cultivated several key trade routes by sea that formed the backbone of their formidable trading empire. These critical maritime corridors enabled Carthage to exchange goods with diverse cultures and accumulate wealth.

Vibrant Mediterranean Trade

The Carthaginians were perfectly positioned to develop active trade with their Mediterranean neighbours. They traded extensively along the North African coast, exchanging metals, livestock, slaves, ivory, and other prized commodities with areas like Numidia and Mauretania. Looking north, they trafficked heavily with the island civilisations of Sicily, Sardinia, Corsica, and the Balearics off the Spanish coast, absorbing agricultural goods like olive oil, wines, and dried fruits in return for Carthaginian crafts and manufactures. They traded with Greek city-states over the sea to the east, like Syracuse, the Levant and Aegean. This Mediterranean trade formed the core of Carthage's commercial activity and allowed diverse cultural influences to mingle.

Remarkable Trans-Saharan Trade

Seeking to access valuable resources from deep inland, the intrepid Carthaginians extended trade routes south across the perilous Sahara Desert. Camel caravans journeyed hundreds of miles into West Africa, connecting Carthage with prosperous centres like Timbuktu. These Sub-Saharan civilisations imported commodities unavailable closer to home, including precious gold, ivory from forest elephants, rare spices and incense like frankincense, exotic animal skins, feathers and live specimens for menageries. This difficult trans-Saharan trade introduced rare luxuries and animals from far-flung interior Africa to Carthaginian markets.

Exploratory Atlantic Trade

In their quest to expand maritime trade, the expert Carthaginian sailors may have voyaged beyond the Strait of Gibraltar into the Atlantic Ocean. Following the northwest African coast, they likely traded with local tribes and possibly even reached the Canary Islands to trade. This demonstrated the Carthaginians' advanced long-distance seafaring knowledge and desire to push trading boundaries further westward. However, firm evidence confirming their Atlantic trade still needs to be discovered.

Carthaginian Port Cities That Enabled Commerce

The Carthaginians established numerous strategic port cities to support their vast maritime trade network stretching across coastal North Africa, the Mediterranean Sea, and the Atlantic Ocean. These ports served as naval bases, government centres, shipyards and thriving commercial hubs that connected Carthaginian merchants with regional and international trade.

The Bustling Harbor City of Carthage

The capital city of Carthage itself was the jewel of the Carthaginian maritime trading empire. Its man-made circular harbour was a monumental feat of engineering, allowing docking space for over 200 ships at once! The inner harbour included shipyards for naval vessels, while the outer quad-shaped merchant harbour had warehouses for housing cargo. The waterfront districts buzzed with traders, artisans, and foreign merchants from across the known world.

Utica - Agricultural Export Center

Founded by Phoenician traders in 1101 BCE, Utica predated Carthage but later became the second city in the Carthaginian realm. Located near the mouth of the Majardah River north of Carthage, it served as an agricultural hub that exported surplus grains, olive oil, dried fruit and other produce from the surrounding fertile plains. Its natural harbour and warehouses supported this trade.

Key Secondary Ports
The Carthaginians also constructed other strategic ports across their sphere of influence, each serving specific functions. Hippo became a naval base. Hadrumetum harboured a war fleet and merchant vessels carrying olive oil and wine. Lilybaeum, Motya and Ibiza also secured Carthaginian control over regional maritime trade.

Carthage's Prime Location as a Major Port City
The city of Carthage enjoyed an advantageous geographic location that enabled it to become a major port and naval power in the ancient Mediterranean world. Situated on the Gulf of Tunis along the North African coast in what is now Tunisia, Carthage occupied a prime position for maritime trade and naval operations.

Carthage had easy access to important sea lanes crisscrossing the Mediterranean, connecting trade partners and colonies scattered across North Africa, Sicily, Sardinia, Spain and beyond. Its double harbour featuring interior and exterior ports could accommodate hundreds of ships simultaneously, handling the bustling trade traffic flowing through the city. The outer harbour's narrow entrance was also easily defensible against naval threats.

Beyond its excellent harbours, Carthage had a spacious, defensible peninsula ideal for a port city. The city sat elevated above the surrounding landscape, providing visibility and protection. Fresh water springs and fertile agricultural lands in the vicinity could support a large urban population. Timber from nearby forests allowed Carthage to build abundant ships to dominate Mediterranean trade and warfare.

Capitalising on these natural advantages, shrewd Carthaginian rulers developed Carthage into the preeminent naval and commercial power in the western Mediterranean for centuries leading up to the Punic Wars. The city's unmatched harbours, combined with its geographic location and resources, enabled it to become a hub of trade and naval might unlike any other in its era. The city's status fuelled Carthage's rise as one of the ancient world's most strategically situated port cities.

Far-Reaching Economic Impacts of Carthaginian Trade
The Carthaginians' vast maritime trade network allowed them to control commerce across the ancient Mediterranean, resulting in tremendous economic expansion and prosperity. Some key effects included:

- Local manufacturing boomed as artisans could readily access imported materials like dyes, metals, and ivory to produce finished goods. Carthage became a production hub exporting textiles, pottery, metalwork, glasswork and other crafts.
- Agricultural cultivation expanded to meet foreign export demand like wine, olive oil, grains, and fruit. Regional food production flourished.
- Wealth poured into Carthage as trade revenues enabled monumental building programs, arts patronage, and a high standard of living for merchant elites.
- Trade policies ensured fairness through official oversight of ports, customs, duties, tariffs, and navy patrols to protect sea lanes. This allowed sustained commerce.
- A standardised currency was established to facilitate trade across Carthage's sphere of influence, simplifying business transactions.

- Piracy, local resistance to Carthaginian dominance, and rivalry with emerging powers like Rome challenged their trading hegemony over time.

Hanno and HIS voyages

The Carthaginian leader Hanno was renowned as a skilled navigator who led daring seafaring expeditions to expand Carthage's reach along the northwest African coast. He was entrusted with a fleet of 60 ships on a mission to explore uncharted territory and establish new colonies.

Venturing beyond the Straits of Gibraltar, Hanno founded seven colonies along the coast of present-day Morocco. But he pressed further on an epic voyage down the Atlantic coast of Africa. During this expedition, he encountered something astounding - a heavily populated island inhabited by wild and aggressive people, the likes of which he had never seen.

When Hanno tried capturing some of the male inhabitants, they violently resisted. These savage males could not be subdued and escaped by clambering up cliffs and hurling stones at the Carthaginians. Hanno's men seized three females but bit and scratched their captors so fiercely that the Carthaginians were forced to kill them. They took the skins of the females back to Carthage as evidence of this exotic encounter.

Hanno dubbed these previously unknown people "gorillae." By some interpretations, these were western gorillas - one of the earliest known records of the great apes. Hanno's concise account states: "The females were much more numerous than the males and had rough skins: our interpreters called them Gorillae. We pursued but could take none of the males. They all escaped, so we took three of the females, but they made such violent struggles, biting and tearing their captors, that we killed them and stripped off the skins."

Just how far down the African coast Hanno ventured remains uncertain. Some scholars believe he sailed as far south as Senegal. Others argue for Gambia or Mount Cameroon based on Hanno's description of a great mountain. Regardless of the final destination, his arduous journey proved that determined Carthaginian seafarers could push beyond the edges of existing maps into thrilling new frontiers. Hanno returned with exotic tales that would stir the imagination of future generations of bold African explorers.

Legacy of Carthaginian Maritime Trade

The Carthaginians built an unparalleled maritime trading empire in the ancient world through pioneering innovations in shipbuilding navigation and establishing a vast network of trade routes and strategically placed port cities. The scale and sophistication of their trading activities transformed economies across the Mediterranean, leaving a legacy that continues to shape global commerce today.

Lessons for modern businesses

Following are the lessons that modern businesses can learn from their explorations of trade routes and colonies.

Strategic Location is Critical

Carthage was situated in an ideal spot for maritime trade and leveraged its geographic location to maximum benefit. Businesses should similarly seek optimal locations to access suppliers, customers, and transport routes. Prime real estate, like ports and trade crossroads, is invaluable.

Leverage Your Strengths

Carthage utilised its excellent harbours, shipbuilding resources, and mercantile skills. Businesses should identify and fully capitalise on core competencies rather than try to be well-rounded. Excel at what you do best.

Build a Vast Network

Carthage established colonies and partnerships across the Mediterranean to extend its trade reach. Businesses must forge networks with suppliers, distribution channels and strategic allies to increase their footprint and influence.

Protect Your Assets

Carthage's navy secured trade routes from pirates and enemies. Companies must safeguard production, transportation and intellectual property from physical and cyber threats. Security enables trade.

Explore New Horizons

Carthaginian voyages expanded knowledge and access to new resources. Firms must keep exploring new technologies, markets, and opportunities for future growth, not cling to legacy businesses.

By applying these lessons from pioneers like Carthage, modern enterprises can establish far-reaching, resilient networks that withstand competition and create lasting value.

In conclusion, the extensive maritime trade networks established by the ancient Carthaginians provided the lifeline that enabled their civilisation to flourish. Strategic ports were hubs for commercial exchange and naval dominance across the Mediterranean. Protected trade routes interwoven throughout Carthaginian territories and colonies fed the empire's prosperity. Although Carthage ultimately fell to Rome, its legacy as one of antiquity's most dominant mercantile powers endures. The ingenuity and entrepreneurial spirit that drove its far-reaching nautical trade pioneered sea routes still traversed by modern ships today. Carthage's seamanship and commerce left an indelible mark on history.

Chapter 7

COINAGE AND MONETARY SYSTEM

The Carthaginian civilisation was prominent in the Mediterranean maritime trade and commercial power. This civilisation had a very sophisticated system of currency and coinage, which played a very important role in the economic and political influence. The Carthaginians inherited the financial system from the Phoenician ancestors, which was known for trade networks and seafaring. The coinage and currency of Carthage were an essential component of the cultural legacy of the ancient civilisation. These coins were a testament to the rich Carthaginian history and the main reason for trade facilitation.

Carthaginian Currency and Coinage

The Carthaginians used a huge variety of metals for coinage. These include gold, silver, and bronze. They could express and refine the metals, allowing them to produce a wide range of coins for different economic transactions. These coins comprise intricate designs and inscriptions. The coins symbolise theological figures and goddesses, which reflect the religious and cultural traditions of the Carthaginian people. The Greeks and Egyptians influenced the Carthaginian currency system. This cultural exchange contributed to the diversity of Carthage and coinage.

They were engaged in countermarking practice, which involves marks or symbols on the existing coins. It was used to verify the authenticity of the coins and currency, which depended on the economic conditions. The coins have also played a significant role in the vast trade network. They were used in commerce across the Mediterranean region. These coins and currency were spread to Sicily, Sardinia, Spain, and North Africa. The wide acceptance of the Carthaginian coinage depicts the economic influence of Carthage.

Impact of the Coinage System on Carthaginian Trade

The coinage system had a very profound impact on the Carthaginian trade. It has contributed to the success and influence of the Carthaginian civilisation. The introduction of the coinage system made the trade in the ancient times more efficient. This is because the merchants and the traders rely on consistent coin denominations. They make transactions very smoother and reduce the need for barter systems. The coinage of Carthage was widely accepted in the Mediterranean region, which has greatly facilitated trade in distant areas. Through this, the traders could easily carry the Carthaginian coins to various markets because this currency was easily accepted as a medium of exchange in distant regions.

These coins have played an important role in integrating Carthage into extensive trade networks. They were used for conducting business within the Carthaginian and other regions. The Carthage was used to exchange goods with the regions, including Sicily, Spain, and North Africa. It was a key player in the Mediterranean trade. These coins were also used to finance the trade ventures. They were used for purchasing goods and paying labour as well. Moreover, the availability of this coin made it easier to assemble the necessary resources for the trading purpose. This coinage system allowed the Carthaginians to compete with their rivals, including Rome.

The traders of the Carthaginians were engaged in commercial activities, contributing to the economic strength. It has also encouraged the maintenance and establishment of great routes and ports. This is because they have invested in infrastructure development, which has supported their trade, including the road network and harbours. It has also influenced and boosted the local cities and the economic structures within those local economies. The presence of the Carthaginian coins has served as the main evidence of the Carthaginian trade. These artefacts have helped archaeologists and historians reconstruct the trade routes and the economic relations between regions.

Monetary Policies and Regulations

Legal regulations and monetary policies are essential tools that central banks and governments use to manage any country's money supply and economic stability. These policies are very helpful in promoting economic growth, controlling inflation, and maintaining financial stability.

The Carthaginians have minted their coins for Commerce and trade throughout the Mediterranean region. These coins were made of silver and were of various designs. Carthage was a major trading power, and its extensive trade networks influenced its monetary policies. They used the stable currency they developed to facilitate commerce across their empire.

The Carthaginians also had a well system with numerous denominations of coins to facilitate the transactions. Like many ancient civilisations, they had smaller denominations for everyday transactions and larger denominations for significant trade throughout the different regions. The economy was heavily reliant on trade. So, the monetary policies of the Carthaginians were designed to support and benefit from their trade.

One of the main objectives of the monetary policy of the modern businesses is to maintain price stability. It can be done easily by controlling the inflation. The central banks usually set an inflation target to ensure that the general level of the prices remains stable. The monetary policies are also designed to support economic growth. The lower interest rates help encourage investment and stimulate economic expansion. In some cases, maintaining the exchange rate stability is crucial for the economy and trade.

Regulations on the Coinage System:

The Carthaginians have established exchange rates for their coins. It was in relation to the commodities or the currencies to facilitate international trade. They have issued various coins with different denominations to support economic transactions. Carthage had a great role as a major trading power in the Mediterranean. So, the currency regulations were very essential for facilitating trade. The regulations were to govern how coins were used in trade practices and commerce. Like many ancient civilisations, Carthage was also involved in taking measures to prevent counterfeiting.

So, the monetary policies and the regulations are very important tools for promoting growth. It also involves a complex rule of the money supply and interest rates, profoundly impacting any country's economic well-being.

Lesson for Modern Businesses in Managing Currencies and Financial Policies

The Carthaginians were known for their advanced financial practices and international trade, which offered detailed lessons for modern businesses in managing financial policies and currency. The following are some of the lessons that must be learned from Carthage to increase your economy and maintain stability.

Currency StandardiSation:

The Carthaginians were known for their standardised and widely accepted coinage system. This was all because of their continuous efforts that made their currency available and trustworthy in distant areas. So modern businesses should also prioritise standardised financial practices, which will help build trust with partners and customers. They should establish consistent invoicing, pricing, and financial reporting practices. It will ensure that the financial transactions are transparent and reliable. This will help in transparent agreements between the partners.

Adaptability Through Countermarking:

The Carthaginians practised countermarking to adjust the coin values, which allowed flexibility. Modern businesses should also be adaptable in their financial strategies. They should respond to the changing customer demands and market conditions. They should adjust the pricing and the payment options to remain competitive in the market. It will also help in accommodating the customer preferences.

Global Trade Expansion:

The Carthaginians were skilled in international trade. They have established a vast trade network, which includes exchanging numerous valuable commodities. Similarly, modern businesses can also learn from them by expanding their global reach. They can easily participate in the international trade networks. Modern businesses should also explore opportunities for entering into new and diverse markets. They should collaborate with International partners, which will help navigate global trade effectively.

Regulatory Compliance:

The Carthaginians have operated within the regulatory framework, which has promoted stability and trust. So, modern businesses also must adhere to the relevant financial regulations. This is crucial to maintain the credibility and also to avoid the legal issues. They should remain updated on the financial regulations and losses, which ensure compliance in all financial reporting and transactions. It will help them in risk management and robust financial governance.

The Carthaginian coinage provides an insight into the ancient civilisation's cultural aspects and economy. These coins featured symbols and motives which emphasised political influence as well. The circulation of the Carthaginian coins highlights the global influence in different regions. These coins serve as historical records that showcase the time's evolving economic and political landscape. So, the modern government should also seek lessons from the Carthaginians for the importance of currency. They should rely on international trade and expand their trade network. Through this, the modern economy can learn the importance of trade for prosperity.

Chapter 8

ECONOMIC IMPACT OF CARTHAGINIAN TRADE

The Carthaginians were known for extensive trade networks in the Mediterranean region. They were located in Tunisia, a major trading power between the 6th and 2nd century BC. The trade of Carthaginians greatly impacted the ancient Mediterranean region by fostering cultural exchange, prosperity, and wealth.

Economic Impact of Carthaginian Trade:

The Carthaginian traders established trade routes that connected the Western and Eastern Mediterranean regions. They have connected with the regions, including Egypt, Greece, Italy, and different parts of North Africa. They were rich in agriculture and resources. The trade involves resources like wine, wheat, barley, metals, olive oil, and other precious metals. This trade has contributed to the economic growth of Carthaginians and the surrounding regions. Their naval power has enabled them to control the sea routes and engage in the maritime trade. The Carthaginians established a powerful navy to protect their trade interest, providing significant revenue.

The trade network of Carthaginians has also facilitated cultural exchange. This is because they have interacted with numerous civilisations and have adopted their spreading of goods, technology, and ideas. So, the wealth generated through the trade has helped the Carthaginians become prosperous. The merchants gathered much wealth, which supported the city's development and military strength.

Economic Prosperity and Consequences

The economic prosperity resulting from the Carthaginian trade had many transformative effects on the individuals in commerce and the city of Carthage. Some of the anecdotes illustrate the transformative effects of the Carthaginians.

Carthaginian Wealth:

The traders and merchants of Carthaginian became the wealthiest individuals in the ancient world. This was all because of their continuous efforts and success in commerce. One of the most famous Carthage traders, named the Hanno, the navigator, gathered immense wealth. He has gathered all this money from his different voyages to the best coast of Africa. They established many colonies and traded numerous valuable resources with African people.

Sea Exploration:

The seafaring of Carthaginians has played an important role in exploring the underwaters. The navigator Hanno led to the discovering of new territories along the African coast. He showcased the adventurous spirit of the Carthaginians and contributed to the geographical knowledge.

Diverse Commodities:

The merchants of Carthaginian were known for their ability to trade a huge variety of goods. They have imported exotic goods like spices, stones, and ivory. They have also exported different products like Tyrian purple dye and pottery. This has all led to the exploration of trade and success in the economy of the Carthaginians.

Consequences of Economic Prosperity

The Carthaginian economic prosperity also had some consequences. Their prosperity and expansion in the Mediterranean region created tensions for the Roman Republic, which led to the Punic Wars. The consequences of these wars were devastating for Carthage, which resulted in the loss of economic power. Moreover, the Carthaginian traders have prospered a lot. However, the wealth was not shared equally among the traders, which led to economic inequality.

Impact of Trading Success on Economic Activities

The Carthaginian trading success had a very profound impact on the economic activities. The strategic location of the Carthaginians in the central Mediterranean region has allowed them to control and benefit from trade routes. This enabled the Carthaginian merchants to engage in the profitable exchange of various goods, including exotic products, textiles, spices, and metals.

The Carthaginian government has recognised the importance of trade. They helped in activities and supported and invested in maintaining and constructing a powerful navy. This was to protect the merchant vessels and secure the trade routes. Additionally, they established favourable trade agreements for the Carthaginians, which were conducive to commerce and innovative trade practices.

They developed a sophisticated system of maritime navigation, which helped them explore new markets and establish trade posts. The use of standardised weights has enhanced the trust in the trade dealings between the partners. So, the economic success of the Carthaginian trade facilitated the cultural exchange and brought new ideas and artistic influences. This exchange has contributed to the enrichment of Carthaginian society.

Key Lessons from Carthaginian Trade

The Carthaginian trade experience offers numerous economic lessons for modern businesses to increase their presence in the global economy.

Innovation in Trade Practices:

The Carthaginian traders were prominent in Maritime navigation and standardised sights and measurements. These innovations were very useful in efficient trade practice and embracing technological advancements. So modern businesses should also apply innovative practices in the trade sector to get benefits and increase their economic growth rate.

Geopolitical Risk:

The Carthaginian economic success and expansion were related to the geopolitical situation. They have highlighted the importance of diversifying trading partners and understanding the geopolitical landscape. So modern businesses should also understand the risk associated with the Geopolitical aspect, which will help them in the future and reduce risk in their business.

Government Support:

The Carthaginian government supported them and invested in the trade, which was instrumental in their success. So, modern businesses should also know the importance of a supportive regulatory environment. They should develop infrastructure that should facilitate economic growth and trade. They should accompany the government officials to support them and build their infrastructure to facilitate trade.

Overall, the Carthaginian trade had numerous economic impacts. Their trade brought a huge wealth to the city. Also, the continuous efforts in economic activity have stimulated growth. They have established numerous trade networks for the exchange of goods. So, modern economies should also learn lessons from them in trading efficiently. They should diversify the trade partners and create innovation in trade practices.

Conclusion

To conclude, Carthage became the wealthiest civilisation for nearly half a millennium until the Roman Empire conquered it. Despite its defeat by the Romans, Carthage they of shipbuilding and trading, which continues to benefit modern engineers and businesses. Today, we have specialised classes and schools for business and shipbuilding. However, the Carthaginians passed down their knowledge from generation to generation, instilling it deeply in their culture, and the new generations learned these remarkable skills from a very early age. As a result, they dominated the Mediterranean trade for over five centuries.

The Carthaginians have left behind valuable lessons in exploration and risk-taking in business, establishing a trading culture, and practising firm and steadfast ethical business principles, which are still relevant to modern businesses. They've also shared their unique shipbuilding techniques through archaeological shipwrecks, and the remnants of their naval power tell stories of their bravery to future generations, illustrating their capability to overcome any threats through their strength. Let's not forget their trading empire, which surpassed the civilisations of their neighbouring regions, leaving a lasting legacy in the world of commerce and trade.

FINAL SUMMARY

Despite being overlooked for most of ancient history, the Carthaginians were a remarkable civilisation that rivalled their Phoenician ancestors.

They were renowned for their naval strength, especially in conflicts with Rome, and their shipbuilding skills that led to famous warships such as the Quinquereme. The foundation behind the vessels was a long and narrow design that made them well-suited for speed and other operations.

Nestled in present-day Tunisia, the city of Carthage enjoyed a coveted perch at the crossroads of the Mediterranean Sea. Possession of vital territories like Sicily, Sardinia, and other Mediterranean islands gave Carthage remarkable success using key trade routes.

They had a sophisticated currency and coinage system, including gold, silver, and bronze. They could express and refine the metals, allowing them to produce a wide range of coins for different economic transactions.

Before the Romans conquered Carthage in about 146 B.C., they left a profound mark on the history still relevant to modern-day businesses.